Organic Chemistry Student's Notebook

For Molecular And Structural Formulas

Christine Dunne

Christine Dunne, Publisher

Salinas, California 2020

Copyright © 2020 Christine Dunne

All Rights Reserved.

ISBN-978-0-578-62430-3

Printed by Lulu Press, Inc. in the United States of America

First Printing, 2020

Christine Dunne, Publisher

P.O. Box 2002

Salinas, California 93902

www.deadland.co

www.ingramcontent.com/pod-product-compliance
Lightning Source LLC
Chambersburg PA
CBHW080343170426
43194CB00014B/2662